The Number Detective

JON MILLINGTON

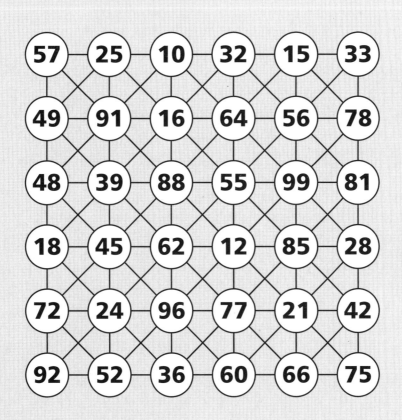

TARQUIN PUBLICATIONS

Hidden Numbers, Detective Skills, Clues, Logical Thinking ...

The answers to all the puzzles in this book are hidden numbers and detective skills are needed to find out what they must be. Each puzzle is liberally sprinkled with clues and you can enjoy all the benefits and delights of logical thinking as you work them out!

While trial and error has its part to play, especially when starting with a new puzzle, the greatest satisfaction comes from using the properties of numbers to avoid unnecessary computation and to jump to the answer in an elegant way. Just like a skilful detective!

Many mathematical words have been used in the puzzles and you might not know all of them. However, there is a glossary on pages 53-56 where you can look up anything that you are not sure about.

On pages 48-49, there are lists of primes, squares, factors etc. You will find them very helpful for certain of the puzzles.

I should like to thank my wife Pat for all her help and support during the preparation of this book.

© 2004   Jon Millington
© 1999   First Edition
I.S.B.N: 1 899618 33 3
Design: Magdalen Bear
Printing: Five Castles Press, Ipswich

Tarquin Publications
Stradbroke
Diss
Norfolk IP21 5JP
England

Lowest number

1. What is the lowest number that has
all three of these properties?

a. It is the sum of five consecutive numbers.

b. It is the sum of two consecutive odd numbers.

c. It is the sum of three consecutive even numbers.

Additions through the grid

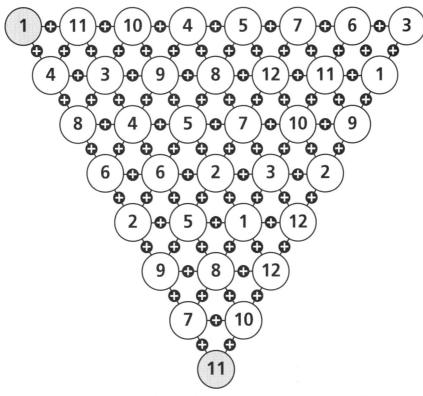

2. Start with 1 in the shaded circle and then work through the
grid so that the running totals are even and odd alternately.
What is the total on the final shaded square?
This must be done without retracing your steps.

Three triples

3 4 5 6 7 8

28 30 35

3. Group these numbers into three groups of three so that the products of each of the triples are the same. What is this product?

Five pairs

4. Pair these ten numbers so that the difference between the two numbers in each pair is exactly divisible by 7.

6 **28** **58** **98**

45

78

83

17 **64** **37**

Which is the divisor?

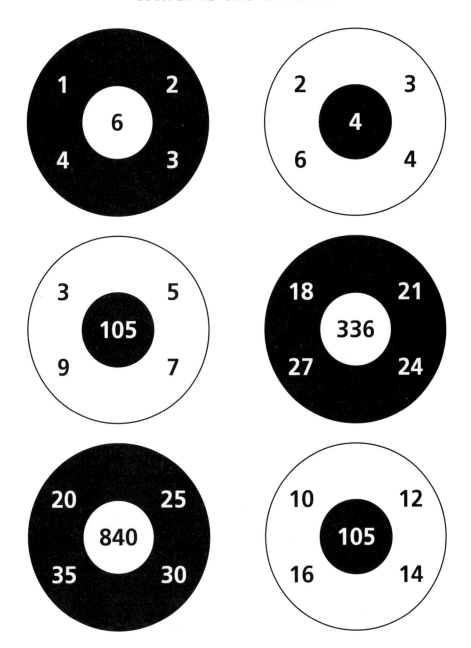

5. The number in the centre of the circle is the result of multiplying three of the numbers in the outer ring and then dividing by the fourth.
Which of the numbers is the divisor in each example?
Try trial and error, but there is a clever way too.

What's next?

120   210   336

6. Each of these three numbers is both the sum of three consecutive numbers and the product of three consecutive numbers. What is the next number in this sequence?

A four figure number

7. When a certain four-digit number is multiplied by 4, its digits appear in reverse order. It also has both of the following properties.

Its first digit is a quarter of the last one.

And its second digit is one less than the first.

What number must it be?

Prime products?

15   35   143
437   323   899

8. Each of these six numbers is the product of two primes. Which is the odd one out and why?

Fly away balloons

9. If a number on any helium balloon is included in the list below, the balloon is released.

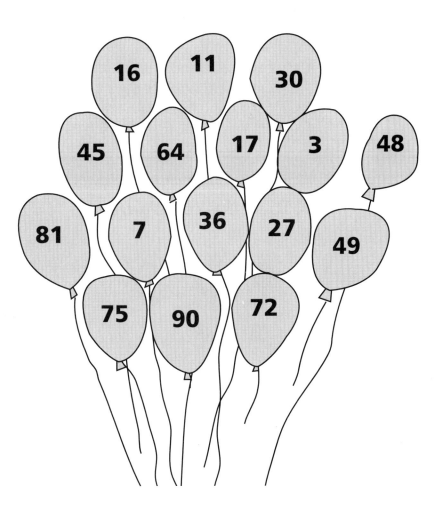

a. Balloons with a multiple of both 3 and 5.

b. Balloons with a square number.

c. Balloons with a prime number.

d. Balloons with numbers having 8 as a factor.

Which balloon remains?

Sets of consecutive numbers

10. Certain sets of three consecutive numbers, whose sum is less than 50, make the following patterns. Which sets of numbers are they?

a. A prime, followed by a cube, followed by a square.

b. A triangular number, followed by a square, followed by a prime.

c. A square, followed by a triangular number, followed by a prime.

d. Multiples of 2, 3 ,4 in this order.

Zero to eight

1	2	0
3	4	
5	6	
7	8	

11. Of the six numbers that can be obtained by adding numbers which come one above another in pairs, there is only one that will not divide exactly into 120. What is it?

A special Fibonacci number

1 1 2 3 5 8 13 21 ??

12. This number is a member of the Fibonacci sequence.

a. It is also a prime less than 1000.

b. Adding 8 to it gives the next largest prime.

What number is it?

A special prime

13. What prime is this?

a. It has two digits and when its digits are reversed, it is also prime.

b. In addition, the sum of the two digits is also prime.

Alphabetical Order

14. If the counting numbers are written out in full and arranged alphabetically, then **FOUR** comes before **SEVEN**, but after **FIVE**.

a. ZERO comes last in this alphabetical arrangement but out of the first hundred numbers, which number comes first?

b. Which comes second?

Missing the point?

15. In all the addition sums below, the decimal points have been omitted but in each case the answer is a square.
Every number lies between 1 and 100.

$$159 + 209 + 251 = a^2$$

$$293 + 453 + 217 = b^2$$

$$326 + 704 + 936 = c^2$$

$$455 + 885 + 965 = d^2$$

a, b, c & d are whole numbers.
What are their values?

Triple sums

16. Divide these nine numbers into three groups of three so that the sum of each group is the same.

11   73

91   35   43

85   63   25   51

Semi-primes

17. Can you find the semi-prime which is

    a. A member of the lowest triple of consecutive semi-primes.

    b. And also the sum of consecutive factorials?

What shape appears?

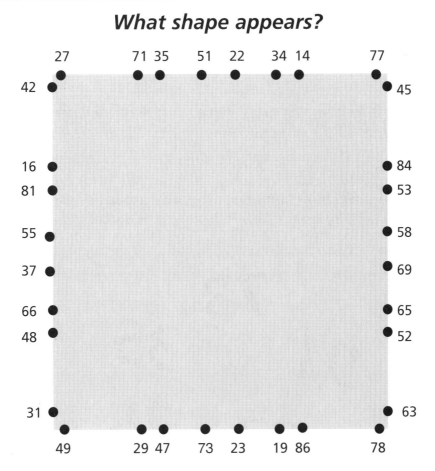

18. Join with straight lines those points whose numbers add up to 100. What shape do the lines envelop?

Missing digits

19. Here are four sums where some of the digits have been replaced with question marks.
Discover what the sums must have been.

a.

```
   ? 5 7
   2 2 ? 6
 + ? 9 4 8
 ─────────
   4 3 2 ?
```

b.

```
   7 ? 1
   1 5 7 ?
 + 1 ? 9 8
 ─────────
   ? 7 0 2
```

c.

```
   1 1 ? 7 1
 - ? 6 ? 9
 ─────────
   8 6 4 ?
```

d.

```
   8 3 ? 0
 - 3 ? 0 ?
 ─────────
   ? 9 3 6
```

There is an ingenious connection between the answers and the missing digits. Can you see what it is?

A special two-digit number

20. Just one two-digit number has this property.

When 2 is subtracted from it
and the result doubled,
it becomes
the same two-digit number
with its digits reversed.

What is it?

Nought to sixty

21. Delete numbers following the rules below until just one number remains.

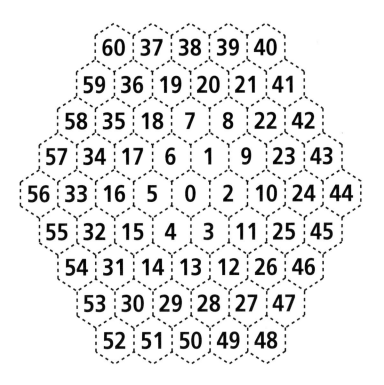

a. Delete all the numbers that are prime or square.

b. Delete all the numbers that differ from a prime or a square by 1.

c. Delete any numbers that are multiples of 5 or differ from a multiple of 5 by 1.

d. Delete any numbers that are multiples of 7 or differ from a multiple of 7 by 1.

Which number remains?

Clock watching

Looking at the clock face, find the following numbers.

22. What number is this?

    a. It is half one of the numbers on the clock face.

    b. It is also two-thirds of another number on the clock face.

    c. And three-quarters of another.

23. What number is this?

    a. On the clock face this number is opposite a prime.

    b. However, it is not itself prime.

    c. The numbers on either side are both composite.

24. What number is this?

    a. It is the sum of every third number shown.

    b. It is a semi-prime.

    c. Its digits are different.

Phone numbers

This is an example of a six digit telephone number made from two groups of three digits separated by a space.

012 345

25. Find the telephone number for which both these statements are true.

a. The final three digits are divisible by 12 and so are the final two digits.

b. It starts with 012 and continues so that each digit is larger than the one before.

26. Here is another telephone number to deduce.

a. The first digit is zero and it is followed by a two digit square number.

b. The whole number reads the same backwards or forwards.

c. The sum of its digits is 20.

Ten factors

27. Before and after 100.

a. Which numbers up to and including 100 have ten proper factors?

b. Which is the first number after 100 which has more than ten proper factors?

Sum and difference

28. What number less than 100 is this?

a. Its digits add up to 12.

b. If the unit digit is taken from the tens digit and doubled the answer is also 12.

Primes from digits

29. This is one of the sets of primes that can be made using each of the digits 1 - 9 once.

In this set no prime has more than two digits.

Only one prime in the set has its digits in ascending order.

What are the primes in the set?

Find the fraction

30. This number is a fraction which will simplify to two-thirds.

a. All digits in the numerator are the same and all digits in the denominator are the same.

b. Adding 34 to both the numerator and the denominator produces a fraction which reduces to thirty-nine fiftieths.

1 4 9 16 ... ... 343 512 729

31. This number is a square less than 1000.
a. In addition, it is a multiple of 7.
b. The sum of its digits is divisible by 9.
What number is it?

Numerators and Denominators

$\frac{1}{1}$	$\frac{1}{2}$	$\frac{1}{3}$	$\frac{1}{4}$	$\frac{1}{5}$	$\frac{1}{6}$
$\frac{2}{1}$	$\frac{2}{2}$	$\frac{2}{3}$	$\frac{2}{4}$	$\frac{2}{5}$	$\frac{2}{6}$
$\frac{3}{1}$	$\frac{3}{2}$	$\frac{3}{3}$	$\frac{3}{4}$	$\frac{3}{5}$	$\frac{3}{6}$
$\frac{4}{1}$	$\frac{4}{2}$	$\frac{4}{3}$	$\frac{4}{4}$	$\frac{4}{5}$	$\frac{4}{6}$
$\frac{5}{1}$	$\frac{5}{2}$	$\frac{5}{3}$	$\frac{5}{4}$	$\frac{5}{5}$	$\frac{5}{6}$
$\frac{6}{1}$	$\frac{6}{2}$	$\frac{6}{3}$	$\frac{6}{4}$	$\frac{6}{5}$	$\frac{6}{6}$

32. Using the numbers 1 to 6 as numerators and denominators gives rise to thirty-six different fractions.
a. How many of these are equivalent to whole numbers?
b. How many are equivalent to halves?
c. How many are improper?

Consecutive number sums

33. Here are three consecutive number totals.

123   1234   12345

a. The number 123 is the sum of three consecutive numbers.
What are they?

b. The number 1234 is the sum of four consecutive numbers.
What are they?

c. The number 12345 is the sum of five consecutive numbers.
What are they?

Number ring

34. This diagram shows the numbers 1 to 12 arranged
in the form of a ring of cells.

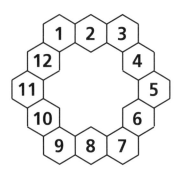

What is the only number that is
both the product of two opposite cells
and also the product of two adjacent cells?

An exact cube root

35. Six numbers, each one larger than the one before, have a total which has an exact cube root.

The first is the next odd number after eight squared.

The second is the next prime.

The third is the next number whose digits add up to 8.

The fourth is the next square.

The fifth is the next number whose digits add up to 4.

The sixth is the next cube.

What is the total and what is the cube root?

Squares and cubes

36. Which of these numbers is the sum of a square and a cube?

150

810

392

252

722

576

686

When divided by the square, only one of these numbers produces a prime. Which?

All different

37. This number consists of three consecutive digits.

When it is added to the other five numbers,
which can be made with these three digits,
the total has all its digits different.

What is it?

A 5 x 5 magic square

38. Complete this magic square which uses each of the numbers
from 1 to 25 once. All its rows, columns and diagonals
must have the same total.

17		1		15
	5		14	16
4		13	20	
10	12	19		
	18		2	

Start by working out
what the total of each of the five rows or columns must be.
It is helpful to know that the sum of the
first 25 natural numbers is 25 x 26 divided by 2.

Equal pair sums

39. The only set of three prime numbers which are consecutive odd numbers are three with single digits.

Make six three digit numbers
by arranging these primes in every possible order.

a. It is a curious fact that these six numbers can be paired off so that the sums of each pair are equal. What is this sum?

b. Does it work for any set of three digits?

Eight and nine and seven

40. This number is divisible by 8 and it is less than the cube of 8.

When 9 is subtracted from it, it is divisible by 9.

When 7 is added to it, it is divisible by 7.

What is it?

Lucky thirteen

41. The arithmetic of this sum is correct but when the numbers are written out as words, one side is not an anagram of the other.

TWO + SEVEN = THREE + SIX

Remarkably,
there is an equation like this where both sides add up to 13 and where each side is an anagram of the other.

Which pairs of numbers have this property?

Star numbers

42. All the numbers at the points of the stars lie
between 200 and 300.
Identify which is the odd one out in each set of five.
The number at the centre gives a clue to what to look for.

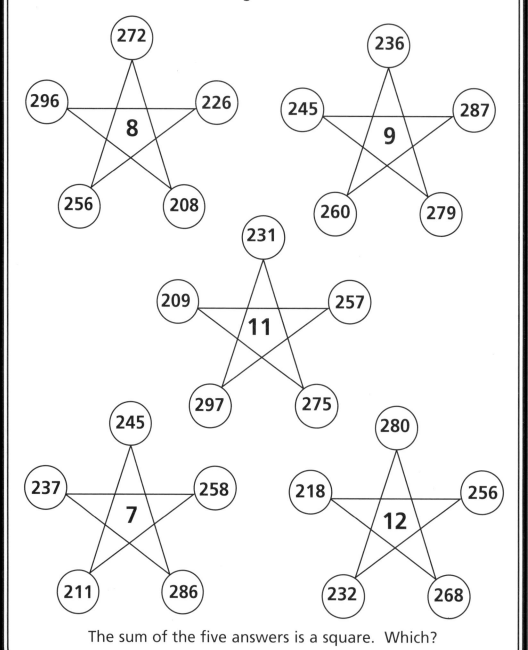

The sum of the five answers is a square. Which?

A four-digit number

????

43. Which four-digit number has all these properties?

a. Its second digit is twice the first.

b. Its fourth digit is three times the third.

c. All its digits are different.

d. No two of its digits are consecutive.

Six consecutive numbers

3 + 4 + 5 = 12
6 + 7 + 8 = 21

44. The six consecutive numbers
3, 4, 5, 6, 7, 8
have the curious property
that the sum of the first three is a two-digit number
and the sum of the last three
has the same two digits, but reversed.

There are two other sets of six consecutive numbers
which have this property.

What are they?

A repeating rule

45. Take any number and if it is even, halve it.
If it is odd, multiply it by 3 and add 1.
Continue the process.
How does it end?

For example

17

is odd, so multiply it by 3 and add 1. This gives

52

which is even, so halve it. This gives

26

which is even, so halve it. This gives

13

which is odd, so multiply it by 3 and add 1. This gives

40

The sequence then continues **20**, **10**, **5**, **16**, **8**, **...**

What happens next? Try starting with other numbers.

Squares in two ways

46. What number must this be?

It is the sum of two squares,
the larger one being the square of 63.

It is also the sum of two different squares,
the larger one being the square of 60.

Factorials

47. This number has a factorial which is less than 100 000.

a. In addition, the digits of the factorial add up to 9.

b. Also, three times the number is another factorial.

What number is it?

Equivalent to a half

$$\frac{21 + ?}{71 + ?}$$

48. The problem is to make this fraction equivalent to a half by adding the same number to both the top and the bottom. In this case, adding 29 will produce the desired result.

$$\frac{21 + 29}{71 + 29} = \frac{50}{100}$$

What number has to be added to both the top and bottom of each of these fractions to make a half?

$$\frac{25 + ?}{61 + ?} \qquad \frac{35 + ?}{83 + ?} \qquad \frac{45 + ?}{105 + ?}$$

What would be the next fraction in this sequence?

Pascal's triangle

<pre>
            1
          1   1
        1   2   1
      1   3   3   1
    1   4   6   4   1
  1   5 10 10   5   1
1   6 15 20 15   6   1
1   7 21 35 35 21   7   1
1   8 28 56 70 56 28   8   1
.   ..   .
</pre>

This well-known pattern starts with a single 1
and then the numbers in each new row are calculated
by simply adding the numbers on the row above
which are just to the left and to the right.

Pascal's triangle continues for ever and
serves as a basis for the next four puzzles.

Pascal puzzle 1

49. What number must this be?

a. It is one of the numbers which only appears three times
in the complete expression of Pascal's triangle.

b. It is a two-digit number which is the product
of consecutive integers.

Pascal puzzle 2

50. This number is one of those displayed opposite.

a. It is in a row whose total is 64.

b. The six numbers surrounding it add up to 97.

What is it?

Pascal puzzle 3

51. Which number in Pascal's triangle has these properties?

a. It lies in a row whose total has three digits.

b. The difference between it and its neighbour is 90.

c. Each digit is smaller than the one before.

Pascal puzzle 4

52. Which number is this?

a. It is the sum of two adjacent numbers displayed opposite.

b. It is divisible by 7.

c. The sum of its digits is 9.

Stepping from cell to cell

From START, step through adjacent cells
obeying the following rules.

When moving across to the right, add the number
in the new cell to the existing number
and when moving left subtract.

When moving upwards, multiply by the number
in the new cell and when moving downwards divide.
(The result of a division must always be a whole number.)

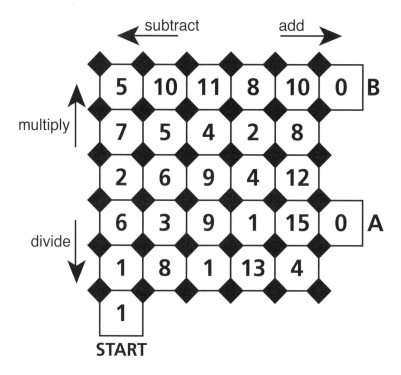

53. What is the total at A if the answer on each cell
is always a square or a cube?

54. What is the total at B if the answer on each cell
is never divisible by 5 or 9?

55. What is the largest total at B that is possible if
no steps to the left or downwards are allowed?

Start with a thousand

56. Start with 1000

Subtract the second prime.
Reverse the digits.
Divide by 17.
Add the first prime.
Take the square root.
Double it.

Subtract 1

57. Start with 1000

Subtract the third
cube number.
Divide by 7.
Reverse the digits.
Divide by 7.
Divide by 7.
Reverse the digits.

Divide by 7.

58. Start with 1000

Take the cube root.
Add 3 raised to the
power of 5.
Divide by 11.
Double it.
Double it and add 1.
Reverse the digits.

Divide by 3.

59. Start with 1000

Add the square of 34.
Divide by 4.
Divide by 7.
Add the square of 9.
Subtract the cube of 4.
Subtract the cube root of 729.

Find the sum of the digits.

Two digit triples

60. Which group of three consecutive two-digit numbers
has the following properties?

a. The sum of all the six digits in the three numbers is 27.

b. The sequence begins with a prime number
but does not end with one.

Four missing digits

61. Find the missing digits
when each of the four sides of this frame
has the same total and all ten digits are different.

Fractions for decimals

0.44   **0.66**

0.22   **0.88**

62. There are many different ways in which these
four decimals could be written as equivalent fractions.

a. What is the smallest denominator that could be the same
for all four fractions
equivalent to these four decimals?

b. And what is the smallest numerator that would be the same
for all four fractions
equivalent to these four decimals?

Both ways round

63. Here are examples of two-digit numbers whose product is the same when the digits are reversed.

$$12 \times 63 = 21 \times 36 = 756$$
$$23 \times 96 = 32 \times 69 = 2208$$

Find the pair of numbers with this property where two of the four multipliers are divisible by 34.

Try investigating further pairs of such reversible numbers.

Numerators and denominators

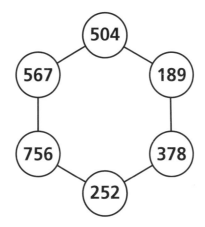

64. By using a pair of adjoining numbers, one as a numerator and one as a denominator, make the following fractions.

a. A fraction equivalent to a half.

b. A fraction equivalent to two-thirds.

c. A fraction equivalent to three-quarters.

Digital keypads

Calculator

7	8	9
4	5	6
1	2	3
0		

This is how the digits 0 to 9 are displayed on a calculator.

65. Look at the display and deduce this number.

a. It is the product of all the digits in the same row or column.

b. It is a multiple of 4 but not of 12.

66. Using the same display, find this number.

a. It is the product of two diagonally adjacent digits.

b. It is the only one of these numbers which is also the product of two other single digits.

Telephone

1	2	3
4	5	6
7	8	9
	0	

The numbers on a telephone are arranged slightly differently from those on a calculator.

67. Using this display, find the number which is the product of all the digits in the same row or column and a multiple of 4 but not of 7.

Common factors

68. Deduce all the common factors
of
the sum of the first seven square numbers
and
the sum of the first ten square numbers.

Start with 1.

Between the prime pairs

69. There are many examples of twin primes.
For instance

17 & 19   71 & 73   419 & 421

The numbers which lie between these pairs,
namely 18, 72, 420 are all multiples of 6.
Is this a coincidence?
What is the largest such number
that is not a multiple of 6?

Prime difference

70. Which two-digit number is this?

a. Between the two digits there is a difference of 4.

b. It differs from a square by 1.

c. It is prime.

Four proper factors

71. There are three pairs of consecutive numbers up to 100 which each have exactly four proper factors.

Of these, the product of only one pair is not divisible by 9.

What is this product?

Three operations

365

953

276

527

111

72. The numbers above are just examples of three-digit numbers which could be used, although any others would do.

Firstly, take any three-digit number and multiply it by 9.

Secondly, reverse the digits of the result of this multiplication and add the two numbers together.

Thirdly, add together the digits in the answer.

What is always true about the result of this sequence of operations and why?

Amicable Maze

73. In this maze it is possible to reach the centre in two ways.

Starting with 1, follow each route,
performing the calculations in order.

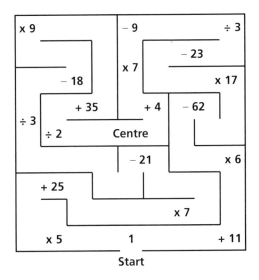

What are the two numbers reached at the centre?
and
why is it called an 'amicable maze'?

Homing in

74. Which number is this?

a. It is a two-digit square number.

b. It has other proper factors in addition to its square root.

c. Those proper factors include at least two numbers
which are consecutive.

Fractions from opposites

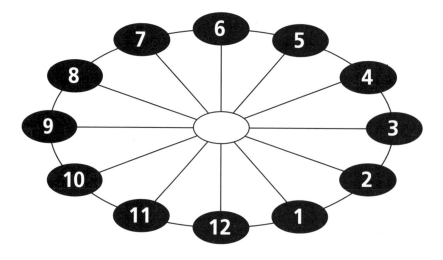

75. What proper fraction is this?

a. It is obtained by dividing a number by its opposite and then reducing it to its lowest terms.

b. It is not a unit fraction.

c. When written as a decimal it does not recur.

Palindromic number

76. The number

1001

is the nearest palindromic number to 1000 which is divisible by 7.

What is the next nearest?

Sets of primes

77. Here are three sets of three-digit prime numbers where each set uses the digits 1 to 9 once.

127 463 859

149 563 827

239 461 587

a. If 659 is the smallest prime in a similar set what are the other two?

b. If 461 is the largest prime in a similar set what are the other two?

Factors and Fibonacci

1, 1, 2, 3, 5, 8, 13, 21, 34, 55, 89

78. The series above shows the Fibonacci numbers which are less than 100.

What is the only number less than 100 which has both these properties?

a. It has exactly six proper factors.

b. Three of those six factors are Fibonacci numbers and the sum of the other three is also a Fibonacci number.

Product pairs

0.4

2.5

0.8

0.625

0.32

3.125

1.25

1.6

79. Pair these numbers so that their products are 1.

Factorial eight

8!

80. Although factorial eight is the product of all the numbers from 1 to 8, it is also divisible by 9 and 10.

Why can you be certain that it will also be divisible by 128?

Missing digits

81. Replace the ?s in each of these multiplication sums. No digit is repeated within each sum.

```
  ? 4 ?              8 ? ?              3 ? ?
    x 6                x 7                x 8
  -------            -------            -------
  ? ? 0              ? 6 ? 4            ? 7 6 0
```

More missing digits

82. Replace the ?s in each of these division sums where there are no remainders.

$$
\begin{array}{r}
2\,?\,4 \\
3\,)\overline{\,?\,9\,?\,}
\end{array}
\qquad
\begin{array}{r}
7\,2\,? \\
7\,)\overline{\,?\,0\,?\,9\,}
\end{array}
\qquad
\begin{array}{r}
6\,?\,6 \\
8\,)\overline{\,?\,0\,0\,?\,}
\end{array}
$$

Triangular numbers

83. All the two-digit triangular numbers are included in this grid. Which of them has no other triangular numbers on any of the four linking lines which pass through it?

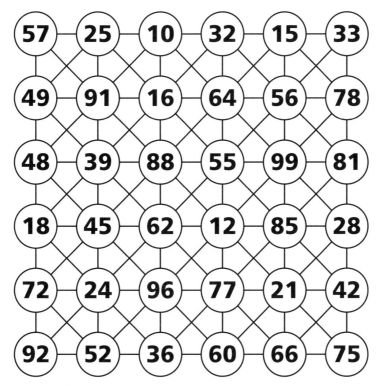

39

Two single-digit numbers

84. Find the only two single-digit numbers
that have this property.

When the greater one is added to the square of the lesser one,
it is exactly a third of the result of
the smaller one being added to the square of the larger one.

Just one out

85. Every digit in each of these sums is just one out!
It might be one greater or one smaller
but the difference is always only 1.
Deduce what the sums should have been.

For example **26 + 57 = 56** is not correct,

but

17 + 48 = 65 is.

a. **739 + 354 = 764**   b. **3778 + 7150 = 8001**

c. **354 - 368 = 313**   d. **536 - 495 = 374**

e. **43 x 2 = 211**   f. **461 x 5 = 4311**

g. **71 ÷ 6 = 25**   h. **685 ÷ 8 = 57**

Double operations

  **-10** **x2**

+4 **÷5**   **÷3** **-9**

86. Mentally place one of these double-operation cards between each of these pairs of numbers so that the calculation becomes correct.

47   = 74

91   = 19

24   = 42

81   = 18

39   = 93

Always one

0.375 $\dfrac{4}{5}$ 0.2 $\dfrac{5}{8}$

$\dfrac{3}{4}$ 0.32 $\dfrac{17}{25}$ 0.25

87. Pair a decimal with a fraction so that their sum is 1.

Domino square

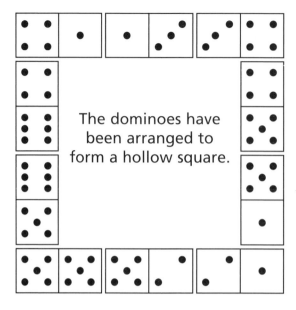

The dominoes have been arranged to form a hollow square.

88. It is easy to see which is the domino with the fewest spots and which is the one with the most spots, but which has the best claim to have an average number of spots?

Digits reverse

89. Surprisingly only one two-digit number meets this requirement.

$$(\blacktriangle \clubsuit + 1) \div 2 = \clubsuit \blacktriangle$$

If you add 1 to it and then divide by 2 , the result is the original number with its digits reversed. What is it?

Common factors

90. Group these nine four-digit numbers into three sets of three, so that each set shares a common factor.

4499

5837

5447

6817

4873

7463

5473

4763

7327

Magic squares

82	68	69	79
71	77	76	74
75	73	72	78
70	80	81	67

91. This is a magic square because all the rows, columns and diagonals add up to the same total.

What number would have to be subtracted from each cell to make a new magic square where all rows, columns and diagonals add up to 42?

The next lowest

2, 6, 10, 14, 18, ...
3, 8, 13, 18, ...
4, 11, 18, 25, ...

92. These three series all go on for ever.
As you can see, each of them includes the number 18.
The puzzle is to work out what is the next lowest number
which appears in all three series.

Decimals or fractions

93. When these grids are completed
they will show the same magic square.
However, one is expressed
in decimals and the other in fractions.

	0.5	
	0.1	

		$\frac{1}{5}$
	$\frac{1}{2}$	

The additional clue that is needed is that
1.5 is the sum of the numbers in each of the
three rows, three columns and two diagonals.

Complete both squares.

Two operations, two orders

Multiply by 4   **Add 8**

94. The two operations above were both carried out
on the same number.
When the operations were performed in one order,
the answer was 140.
When they were carried out in the other order,
the answer was 116.
What was the starting number?

Divide by 4   **Subtract 8**

95. The two operations above were both carried out
on the same number.
When the operations were performed in one order,
the answer was 48.
When they were carried out in the other order,
the answer was 54.
What was the starting number?

Fractions as decimals

$$\frac{1}{2} \quad \frac{1}{3} \quad \frac{1}{4} \quad \frac{1}{5} \quad \frac{1}{6} \quad \frac{1}{7} \quad \frac{1}{8} \quad \frac{1}{9} \quad \frac{1}{10} \quad \frac{1}{11} \quad \frac{1}{12}$$

96. When these unit fractions are converted to decimals,
some have just one figure after the decimal point,
some have two and some have three.
The others are recurring decimals.
Which takes more than three digits before it repeats?

Reciprocal pairs

$$\frac{231}{264}$$ $$\frac{264}{286}$$ $$\frac{385}{315}$$ $$\frac{315}{252}$$

$$\frac{351}{324}$$ $$\frac{324}{396}$$ $$\frac{320}{280}$$ $$\frac{256}{320}$$

97. These fractions can be put into pairs where one is the reciprocal of the other. Match them.

Matching solutions

(85 x 9) + 12

(19 x 12) + 4

(47 x 12) + 3

(55 + 3) x 4

(78 x 8) + 8

(48 x 13) + 8

(70 x 8) + 7

(55 + 56) x 7

98. Match these eight sums into the four pairs which have the same solutions.

What power?

2 x 2 x 2 x   + 3 x 3 x 3 x

99. Choose different powers of 2 and 3,
so that the total comes to each of the numbers below.

7 13 17 31 89 91

Perfect squares

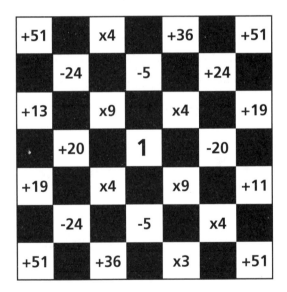

100. The puzzle is to start with the 1 at the centre of
this 7x7 square and find a route through it
which reaches 100 in exactly seven steps.
Each step must be from one white square to a
neighbouring one and on entering each square
the given operation must be performed.
Every number on the route of seven steps is a square.
Can you find this route?

Prime Numbers

2	101	211	307	401	503	601	701	809	907
3	103	223	311	409	509	607	709	811	911
5	107	227	313	419	521	613	719	821	919
7	109	229	317	421	523	617	727	823	929
11	113	233	331	431	541	619	733	827	937
13	127	239	337	433	547	631	739	829	941
17	131	241	347	439	557	641	743	839	947
19	137	251	349	443	563	643	751	853	953
23	139	257	353	449	569	647	757	857	967
29	149	263	359	457	571	653	761	859	971
31	151	269	367	461	577	659	769	863	977
37	157	271	373	463	587	661	773	877	983
41	163	277	379	467	593	673	787	881	991
43	167	281	383	479	599	677	797	883	997
47	173	283	389	487		683		887	
53	179	293	397	491		691			
59	181			499					
61	191								
67	193								
71	197								
73	199								
79									
83									
89									
97									

Squares & Cubes

	Square	Cube		Square	Cube
1	1	1	21	441	9261
2	4	8	22	484	10648
3	9	27	23	529	12167
4	16	64	24	576	13824
5	25	125	25	625	15625
6	36	216	26	676	17576
7	49	343	27	729	19683
8	64	512	28	784	21952
9	81	729	29	841	24389
10	100	1000	30	900	27000
11	121	1331	31	961	29791
12	144	1728	32	1042	32768
13	169	2197	33	1089	35937
14	196	2744	34	1156	39304
15	225	3375	35	1225	42875
16	256	4096	36	1296	46656
17	289	4913	37	1369	50653
18	324	5832	38	1444	54872
19	361	6859	39	1521	59319
20	400	8000	40	1600	64000

Triangular Numbers

1	190	703
3	210	741
6	231	780
10	253	820
15	276	861
21	300	903
28	325	946
36	351	990
45	378	
55	406	
66	435	
78	465	
91	496	
105	528	
120	561	
136	595	
153	630	
171	666	

Proper Factors

4	2
6	2 3
8	2 4
9	3
10	2 5
12	2 3 4 6
14	2 7
15	3 5
16	2 4 8
18	2 3 6 9
20	2 4 5 10
21	3 7
22	2 11
24	2 3 4 6 8 12
25	5
26	2 13
27	3 9
28	2 4 7 14
30	2 3 5 6 10 15
32	2 4 8 16
33	3 11
34	2 17
35	5 7
36	2 3 4 6 9 12 18
38	2 19
39	3 13
40	2 4 5 8 10 20
42	2 3 6 7 14 21
44	2 4 11 22
45	3 5 9 15
46	2 23
48	2 3 4 6 8 12 16 24
49	7
50	2 5 10 25
51	3 17
52	2 4 13 26
54	2 3 6 9 18 27
55	5 11
56	2 4 7 8 14 28
57	3 19
58	2 29
60	2 3 4 5 6 10 12 15 20 30
62	2 31
63	3 7 9 21
64	2 4 8 16 32
65	5 13
66	2 3 6 11 22 33
68	2 4 17 34
69	3 23
70	2 5 7 10 14 35
72	2 3 4 6 8 9 12 18 24 36
74	2 37
75	3 5 15 25
76	2 4 19 38
77	7 11
78	2 3 6 13 26 39
80	2 4 5 8 10 16 20 40
81	3 9 27
82	2 41
84	2 3 4 6 7 12 14 21 28 42
85	5 17
86	2 43
87	3 29
88	2 4 8 11 22 44
90	2 3 5 6 9 10 15 18 30 45
91	7 13
92	2 4 23 46
93	3 31
94	2 47
95	5 19
96	2 3 4 6 8 12 16 24 32 48
98	2 7 14 49
99	3 9 11 33
100	2 4 5 10 20 25 50

Factorials

1!	1
2!	2
3!	6
4!	24
5!	120
6!	720
7!	5040
8!	40 320
9!	362 880
10!	39 916 800

Fibonacci Numbers

1	89
1	144
2	233
3	377
5	610
8	987
13	1597
21	2584
34	4181
55	6765

Solutions to Puzzles 1 to 20

1. a. **60** = 10 + 11 + 12 + 13 +14
   b. **60** = 29 + 31
   c. **60** = 18 + 20 + 22
The sum of a set of consecutive numbers is given by the middle number multiplied by the number of numbers there are.

2. 1 + 11 + 3 + 9 + 5 + 7 + 3 + 1 + 5 + 9 + 7 + 11 = **72**
Always take the option of the odd number.

3. **3** x 8 x **35** = **840**
   4 x **7** x **30** = **840**
   5 x 6 x **28** = **840**
It is a great help to start by finding the prime factors of all the original numbers. Then make each set include the same primes (2,2,2,3,5,7).

4. 98 − **28** = 70, **83** − **6** = 77, **78** − **64** = 14,
   58 − **37** = 21, **45** − **17** = 28
Trial and error is the best approach.

5. 1 x 3 x 4 = 12 and 12 ÷ **2** = 6
   2 x 3 x 4 = 24 and 24 ÷ **6** = 4,
   5 x 7 x 9 = 315 and 315 ÷ **3** = 105,
   18 x 21 x 24 = 9072 and 9072 ÷ **27** = 336
   20 x 30 x 35 = 21000 and 21000 ÷ **25** = 840
   10 x 12 x 14 = 1680 and 1680 ÷ **16** = 105
Find the prime factors of the central number and then decide which of the other numbers have to be multipliers or divisors.

6. 120 = 39 + 40 + 41 = 4 x 5 x 6
   210 = 69 + 70 + 71 = 5 x 6 x 7
   336 = 111 + 112 + 113 = 6 x 7 x 8
   504 = 167 + 168 + 169 = 7 x 8 x 9
To find the three consecutive numbers whose sum is known, divide by 3. To find the three consecutive numbers whose product is known, look for the nearest cube and use its root. In both cases this gives the middle number.

7. **2178**
The number could be 10?4 or 21?8, but attempting to multiply by 4 eliminates the first possibility. Trial and error finds that it must be 2178 because 2178 x 4 = 8712.

8. 15 = 3 x 5   35 = 5 x 7
   143 = 11 x 13   323 = 17 x 19
   437 = 19 x 23   899 = 29 x 31
437 is the odd one out because its proper factors are not twin primes.

9. a. eliminates 30, 45, 75, 90
   b. eliminates 16, 36, 49, 64, 81
   c. eliminates 3, 7, 11, 17
   d. eliminates 48, 72
      27 remains

10. a. **7, 8, 9**
   b. **3, 4, 5** or **15, 16, 17**
   c. **9, 10, 11**
   d. **14, 15, 16**

11. **14**
The possibilities are 4, 8, 12, 6, 10, 14 and of these only 14 does not divide exactly into 120.

12. a. 2, 3, 5, 13, 89, 233
   b. **89** Next prime 97.

13. a. 11, 13, 31, 17, 71, 37, 73, 79, 97
   b. **11** The sum of the digits has to be even and the only even prime is 2.

14. a. **EIGHT**   b. **EIGHTEEN**

15. a. **1.59 + 20.9 + 2.51 = 25**
   b. **29.3 + 4.53 + 2.17 = 36**
   c. **32.6 + 7.04 + 9.36 = 49**
   d. **45.5 + 8.85 + 9.65 = 64**
Write the alternatives for each of the three numbers. For example 1.59 or 15.9 & 2.09 or 20.9 & 2.51 or 25.1. The answers are whole numbers and so it is relatively easy to pick out which one of each pair is needed.

16. **11 + 63 + 85** = 159   **25 + 43 + 91** = 159
   35 + 51 + 73 = 159
All nine numbers add up to 477 and so each triple must add up to 159.

17. a. 33, 34, 35
   b. **33** = 1! + 2! + 3! + 4!

18. All the lines are tangents to a **circle**.

19. a. **157 + 2216 + 1948 = 4321**
   b. **731 + 1573 + 1398 = 3702**
   c. **11271 − 2629 = 8642**
   d. **8340 − 3404 = 4936**
In addition to the repeating digits:
4321 x **1** = 4321   1234 x **3** = 3702
4321 x **2** = 8642   1234 x **4** = 4936

20. **49** (49 − 2 = 47 x 2 = 94)
Use trial and error or algebra.

21. After a. & b. the remaining numbers are
21, 27, 33, 34, 39, 45, 51, 55, 56, 57
c. 27, 33, 57
d. **33**

22. Only **6** satisfies all the conditions.

23. a. 8, 9, 11, 1, 5
b. 1, 8, 9
c. **9**

24. a. 1 + 4 + 7 + 10 = 22 also 26 and 30
b. 22, 26
c. **26**

25. a. ??? 300, ??? 312, ??? 324 etc. to 396
??? 600, ??? 612, ??? 624 etc. to 696
??? 900, ??? 912, ??? 924 etc. to 996
b. **012 348**

26. a. 016 ???, 025 ???, 036 ???, 049 ???,
064 ???, 081 ???
b. & c. Half the digits must add up to 10
Hence **064 460**

27. a. **60** (2, 3, 4, 5, 6, 10, 12, 15, 20, 30)
72 (2, 3, 4, 6, 8, 9, 12, 18, 24, 36)
84 (2, 3, 4, 6, 7, 12, 14, 21, 28, 42)
90 (2, 3, 5, 6, 9, 10, 15, 18, 30, 45)
96 (2, 3, 4, 6, 8, 12, 16, 24, 32, 48)
b. **120** has fourteen factors.
2, 3, 4, 5, 6, 8, 10, 12, 15, 20, 24, 30, 40, 60

28. a. 39, 48, 57, 66, 75, 84, 93
b. **93**

29. 89, 61, 43, 7, 5, 2
Note that 8, 6 and 4 can only be tens digits,
giving 83 or 89, 61 or 67 and 41, 43 or 47.
2 and 5 can only be units digits if they are on
their own.

30. a. $^2/_3$, $^4/_6$, $^6/_9$ or $^{22}/_{33}$, $^{44}/_{66}$, $^{66}/_{99}$ or
$^{222}/_{333}$, $^{444}/_{666}$, $^{666}/_{999}$ etc.
b. By trial $^{(44+34)}/_{(66+34)} = ^{78}/_{100}$
Hence $^{44}/_{66}$

31. a. 49, 196, 441, 784
b. **441**

32. a. **14** $^1/_1$, $^2/_1$, $^2/_2$, $^3/_1$, $^3/_3$, $^4/_1$, $^4/_2$, $^4/_4$,
$^5/_1$, $^5/_5$, $^6/_1$, $^6/_2$, $^6/_3$, $^6/_6$
b. **3** $^1/_2$, $^2/_4$, $^3/_6$
c. There are **15** 'top-heavy' fractions.

33. a. **40, 41, 42**
b. **307, 308, 309, 310**
c. **2467, 2468, 2469, 2470, 2471**

34. Products of opposites: 7, 16, 27, 40, 55, 72
72 is the only product of adjacents (8, 9)

35. 65 + 67 + 71 + 81 + 103 + 125
512, whose cube root is **8**.

36. **150** = 25(5x5) + 125(5x5x5)
252 = 36(6x6) + 216(6x6x6)
392 = 49(7x7) + 343(7x7x7)
576 = 64(8x8) + 512(8x8x8)
810 = 81(9x9) + 729(9x9x9)
252 ÷ 36 = 7, a prime number
Use trial and error, starting with a low number,
as the totals are not very large, or look at the
number tables on page 48.

37. **789** This gives the total 5328.
The other totals are 1332, 1998, 2664, 3330,
3996, and 4662.

38. Row 1: 17, **24**, 1, **8**, 15
Row 2: **23**, 5, **7**, 14, 16
Row 3: 4, **6**, 13, 20, **22**
Row 4: 10, 12, 19, **21**, **3**
Row 5: **11**, 18, **25**, 2, **9**
Overall total is 325 so each row and column
must add up to 65.

39. a. Digits 3, 5, 7 give 357,375,537,573,735,753
357 + 753 = 1110
375 + 735 = 1110
537 + 573 = 1110
b. No, they must be in sequence, eg 1,4,7

40. **504** = 8 x 9 x 7
The number must be a multiple of 8, 9 & 7.
Any multiple of 504 meets the requirements but
only 504 is less than 512, the cube of 8.

41. ONE + TWELVE = TWO + ELEVEN

42. a. **226** is not divisible by 8
b. **279** is divisible by 9
c. **257** is not divisible by 11
d. **245** is divisible by 7
e. **218** has remainder 2, others 4
1225 is the square of 35

43. a. **12??, 24??, 36??, 48??**
b. **??13, ??26, ??39**
c. 1239, 2413, 2439, 4813, 4826, 4839
d. **4826**

Solutions to Puzzles 44 to 70

44. $14 + 15 + 16 = 45$, $\quad 17 + 18 + 19 = 54$
$25 + 26 + 27 = 78$, $\quad 28 + 29 + 30 = 87$
The two sums must differ by 9 and for the total to be the sum of three consecutive numbers, it must be divisible by 3.

45. As soon as the sequence reaches 4, then it begins to cycle the loop **4, 2, 1** for ever. $(1 \times 3)+1= 4$, $\quad 4 \div 2 = 2$, $\quad 2 \div 1 = 1$

46. **4225** (interestingly the square of 65) Use trial and error or the difference of two squares (369) and algebra.

47. a. 6, 7, 8 (6!=720, 7!=5040, 8!=40 320)
b. **8** as $8 \times 3 = 24$ and $24 = 4!$

48. **11**, **13**, **15** and the sequence continues to $(55+17)/(127+17)$
Use algebra or spot the pattern. For example $61 - (2 \times 25) = 11$ and so on.

49. a. 6, 20, 70, 252, 924, and so on
b. **20** = 4×5

50. a. 1, 6, 15, 20, 15, 6, 1
b. **15** $(5 + 10 + 20 + 35 + 21 + 6 = 97)$

51. a. row 8 (128), row 9 (256), row 10 (512)
b. 120, 210 in row 10
c. **210**

52. **126**
Divisible by 7 and 9, hence a multiple of 63. Try 63, 126, 189, etc. Hence 126 (56 + 70).

53. **64** 1, 9, 27, 36, 36, 49, 49, 64

54. **1041** 1, 6, 12, 84, 89, 93, 1023, 1031, 1041

55. **388800** $(1+8+1+13+4) \times 15 \times 12 \times 8 \times 10$

56. 1000, 997, 799, 47, 49, 7, 14, **13**

57. 1000, 973, 139, 931, 133, 19, 91, **13**

58. 1000, 10, 253, 23, 46, 93, 39, **13**

59. 1000, 2156, 539, 77, 158, 94, 85, **13**

60. a. 17,18,19 26,27,28 35,36,37 44,45,46
53,54,55 62,63,64 71,72,73 80,81,82
The middle number must be a multiple of 9.
b. **53, 54, 55**

61. Reading across 5 0 **9 1** **7** 8 3 2 4 6
The sum of digits 0 - 9 is 45, and the total of the top and bottom rows must be even.
Thus the digit missing in the left hand column must be odd.

62. a. **50** $11/50$, $22/50$, $33/50$, $44/50$
b. **66** $66/300$, $66/150$, $66/100$, $66/75$

63. **34 × 86 = 68 × 43**
There are more answers than one might expect and it makes an interesting investigation to find them. The equation $(10a + b) \times (10c +d) = (10b + a) \times (10d +c)$ leads to the result that $a \times c = b \times d$. The digits which will work can be found by looking for numbers which have two pairs of single digit factors. Two possibilities are $6 = 1 \times 6 = 2 \times 3$ and $18 = 2 \times 9 = 3 \times 6$. In all, the numbers are 4, 6, 8, 9,12,16, 18, 24 & 36 giving nine combinations of pairs with this property.

64. a. $189/378$
b. $252/378$
c. $567/756$

65. a. 504, 120, 6, 28, 0, 0, 80, 162
b. **80**

66. a. 35, 48, 45, 32, 8, 15, 12, 5, 0
b. **12** It is equal to 3x4 in addition to 2x6.

67. **120** The row must include either 4 or 8 but not 7 or 0, Hence $4 \times 5 \times 6 = 120$

68. **1, 5, 7, 35**
The sums are 140 & 385. There is a formula for calculating the sum of n square numbers starting with 1. It is $n(n+1)(2n + 1)/6$. This gives some factors directly.

69. **No, it is not a coincidence.** Because the prime pair and the number between them forms a set of three consecutive numbers. Out of any such set, one must be divisible by three and it cannot be either of the primes. Also the middle number must be even, hence a multiple of 6. The only exception is **4** and this must be the largest. There can be no other.

70. a. 15, 26, 37, 48, 51, 59, 62, 73, 84, 95
b. 15, 26, 37, 48
c. **37**

71. a. 44 & 45   75 & 76   98 & 99
b. **5700** = 75 & 76
Neither 75 nor 76 has 9 as a factor.

72. The digit sum is a multiple of 9.
The sum of the digits in any multiple of 9 is always a multiple of 9. So when a multiple of 9 is reversed it is still a multiple of 9. For example 276 x 9 = 2484. Then 2484 + 4842 = 7326 and 7 + 3 + 2 + 6 = 18, a multiple of 9.

73.   220 & **284** are amicable numbers.

74. a.16, 25, 36, 49, 64, 81
b. 16, 36, 64, 81
c. **36** has consecutive factors (2, 3, 4)

75. a. $^1/_2$, $^5/_{11}$, $^2/_5$, $^1/_3$, $^1/_4$, $^1/_7$
b. $^5/_{11}$, $^2/_5$
c. **$^2/_5$** ($^5/_{11}$ = 0.45 recurring)

76.   959

77. a. 659, **743, 821**   b. **257, 389**, 461

78. a. 24, 30, 40, 42, 54, 56, 66, 70, 78, 88
b. **40** (4 + 10 + 20 = 34)
Only 24, 30, 40, 42, 78 have three factors which are Fibonacci numbers.

79.   0.8 x **1.25, 0.4 x 2.5,
0.32** x **3.125, 0.625** x **1.6**

80. Because 128 = **2** x **2** x **2** x **2** x **2** x **2** x **2**
and 8! = 1 x **2** x 3 x **4** x 5 x **6** x 7 x **8**
which has seven factors of **2** from 2, 4, 6, 8

81. a. 145 x 6 = **870**
b. 802 x 7 = **5614**
c. 345 x 8 = **2760**

82. (a). **792** ÷ 3 = 264
(b). **5089** ÷ 7 = 727
(c). **5008** ÷ 8 = 626

83. 55 The two-digit triangular numbers are
10, 15, 21, 28, 36, 45, 55, 66, 78, 91.

84.   2 & **5** (2 x 2) + 5 = 9   9 x 3 = (5 x 5) + 2
Systematic trial and error.

85. a. **628** + **245** = **873** b. **2869** + **6241** = **9110**
c. **463** - **259** = **204**   d. **647** - **384** = **263**
e. **34** x **3** = **102**   f. **570** x **6** = **3420**
g. **80** ÷ **5** = **16**   h. **594** ÷ **9** = **66**

86. (47 **-10**) **x 2** = 74, (91 **+ 4**) **÷ 5** =19,
(24 **x 2**) **- 6** = 42, (81 **÷ 3**) **- 9** =18,
(39 **- 8**) **x 3** = 93

87. **0.2** + $^4/_5$ = 1   **0.375** + $^5/_8$ = 1
0.25 + $^3/_4$ = 1   **0.32** + $^{17}/_{25}$ = 1

88. 4:3
72 spots on 10 dominoes. Average 7.2.
Hence either 4:3 or 5:2.
But 72 spots on 20 halves suggests 4:3

89.   73 + 1 = 74 and 74 ÷ 2 = 37

90. 4499, 4873 & 4763 have 11 as a factor
5447, 5837 & 5473 have 13 as a factor
6817, 7327 & 7463 have 17 as a factor

91. 64

92. 158
Because the differences are 4, 5 and 7
the required number must be 18 + (4 x 5 x7)

93. 0.4 0.9 0.2   0.3 0.5 0.7   0.8 0.1 0.6
$^2/_5$ $^9/_{10}$ $^1/_5$ $^3/_{10}$ $^1/_2$ $^7/_{10}$ $^4/_5$ $^1/_{10}$ $^3/_5$

94. (**27** x 4) + 8 = 116 and (**27** + 8) x 4 = 140

95. (**224** ÷ 4) − 8 = 48 and (**224** − 8) ÷ 4 = 54

96. 0.3333... , 0.6666 ... , 0.1428571... ,
0.1111... , 0.0909... , 0.0833...
Answer **$^1/_3$**

97. $^{256}/_{320}$ ($^4/_5$) and $^{315}/_{252}$ ($^5/_4$)
$^{231}/_{264}$ ($^7/_8$) and $^{320}/_{280}$ ($^8/_7$)
$^{324}/_{396}$ ($^9/_{11}$) and $^{385}/_{315}$ ($^{11}/_9$)
$^{264}/_{286}$ ($^{12}/_{13}$) and $^{351}/_{324}$ ($^{13}/_{12}$)

98. (85 x 9) + 12 = 777 = **(55 + 56) x 7
(19 x 12) + 4** = 232 = **(55 + 3) x 4
(70 x 8) + 7** = 567 = **(47 x 12) + 3
(78 x 8) + 8** = 632 = **(48 x 13) + 8**

99. 7 = 2^2+3   13 = 2^2+3^2 17 = 2^3+3^2
31 = 2^2+3^3 89 = 2^3+3^4 91 = 2^6+3^3

100. 1 (**x9**) = 9, 9 (**-5**) = 4, 4 (**x4**) = 16,
16 (**+20**) = 36, 36 (**+13**) = 49,
49 (**-24**) = 25, 25 (**x4**) = 100
Start with the x9 square below right of the 1.

Glossary

Addition is the process of finding the total of two or more numbers. It is indicated by the '+' sign, which is read as 'plus'. For example, 7 + 4 + 5 = 16 means that the total of 7, 4 and 5 is 16. Subtraction is the opposite process.

Amicable Numbers are pairs of numbers where all the factors of one (except the number itself) add up to the other. The smallest pair of amicable numbers is 220 and 284, because all the factors of 220 (1, 2, 4, 5, 10, 11, 20, 22, 44, 55 and 110, but not 220) total 284 and all the factors of 284 (1, 2, 4, 7, 71 and 142, but not 284) total 220.

Average is the popular name for the result of dividing the total of a group of numbers by however many numbers there are in that group. So the average of 18, 14, 21 and 15 is 17 because 18 +14 + 21 + 15 = 68 and 68 ÷ 4 = 17. Ideally 'arithmetic mean' should be used instead of 'average'.

Composite Numbers are numbers which can be expressed as the product of two or more factors other than 1. So 30 is composite, because 5 x 6 = 30. A composite number can also be called a rectangular number.

Consecutive means following one after another, as in 7, 8, 9, 10 or Tuesday, Wednesday, Thursday.

Cube   A cube, or perfect cube, is obtained by multiplying a number by itself three times. For example, 125 is a cube because it is 5 x 5 x 5. This can be shortened to 5^3 which is read as five cubed or five to the power three.

Decimal Fraction is a fraction made up of one or more denominators which are powers of 10. For example 0.7 means 7/10 and 0.69 means 6/10 + 9/100

Decimal Point is a dot which separates an integer from a decimal fraction, as in 29.4. We write it level with the bottom of the number.

Denominator is the bottom number of a fraction and shows how many equal parts the whole has been divided into. In the fraction 3/4, the denominator is 4.

Difference is the amount by which the larger of two numbers is greater than the smaller. For example, the difference between 5 and 13 is 8.

Digits are the single-figure whole numbers 0, 1, 2, 3, 4, 5, 6, 7, 8, 9. An example of a two-digit number is 36, while 286 is a three-digit number.

Divisible means 'can be divided exactly by'. For example, 14 is divisible by 7.

Division is the process of dividing a number into a given number of parts. See divisor and quotient. It is the opposite process to multiplication.

Divisor is a number that is divided into another, which is called the dividend. The number of times the divisor is contained in the dividend is called the quotient, and anything left over is called the remainder. For example, in 17 ÷ 3 = 5 rem 2, 17 is the dividend, 3 the divisor, 5 the quotient and 2 the remainder. In a division where there is no remainder, both the divisor and quotient will be factors of the dividend.

Equals, indicated by the '=' sign, shows that the amounts on either side of the sign are the same.

Equivalent Fractions are equal to each other, so 4/5 and 8/10 are equivalent fractions.

Even Numbers are divisible by 2, and are all those numbers ending in 0, 2, 4, 6 or 8.

Factor   Any number which can be divided exactly into another is a factor of that number. This includes 1 and the number itself. Thus, the factors of 6 are 1, 2, 3 and 6. Except for squares, all numbers have an even number of factors.

Factorial of a number is obtained by multiplying it by all the whole numbers which are smaller than it. For example, 5 x 4 x 3 x 2 x 1 (which equals 120) is 5 factorial, abbreviated to 5!

Fibonacci Sequence   A famous number sequence: 1, 1, 2, 3, 5, 8, 13, and so on, where each term is the sum of the previous two.

Fraction, also called a proper fraction, is part of a whole. For example, 3/4 is a fraction which shows that a whole has been divided into 4 equal parts of which we have 3.

Improper Fraction   Sometimes colloquially known as a top heavy fraction, has the top number larger than the bottom, for example 7/4.

Integers are positive or negative numbers. Examples of integers are: 24, -3, 250, -79.

Lowest Terms Where the answer to a problem is a fraction, the top and bottom numbers should be the lowest possible. So, the result of adding 1/10 and 7/10, which is 8/10, would be reduced to its lowest terms by dividing both top and bottom by the same number, in this case 2, giving 4/5.

Magic Square is an arrangement of numbers in a square in such a way that the numbers in each row, column and main diagonals add up to the same total, sometimes known as the magic number or magic constant. The smallest magic square contains nine numbers arranged in three rows with three in each row.

2	9	4
7	5	3
6	1	8

Minus is the name given to the '–' sign.

Mixed Number is a whole number followed by a fraction. For example, 5 3/4 is a mixed number. (Five and three-quarters)

Multiple is the product of a certain number and another number. For example, the first five multiples of 6 are 6, 12, 18, 24 and 30.

Multiplication is the process of repeatedly adding a given number a certain number of times. It is indicated by the 'x' sign, which is read as 'times'. For example, 9 x 4 = 36 means that 36 is the result of adding 4 nines or 9 fours. Division is the opposite process.

Natural Numbers are our ordinary counting numbers 1, 2, 3, 4 and so on. They are the same as positive integers.

Negative Numbers, preceded by the '–' sign, are less than zero. For example –8, –1/2, –0.4.

Number Sequence is a series of numbers which are connected by some rule. One example is 2, 5, 8, 11, 14 and so on, where the difference between consecutive terms is 3.

Numerals are the symbols we use to represent numbers.

Numerator is the top number of a fraction and shows how many equal parts of the whole are required. In the fraction 3/4, the numerator is 3.

Odd Numbers are all the numbers which are not even. They leave a remainder of 1 when divided by 2. All odd numbers end in 1, 3, 5, 7 or 9.

Palindromic Numbers are numbers which read the same backwards or forwards. For example 737, 404, 2772, 35853.

Percent is a number out of 100, in other words a fraction where the denominator is 100. The sign for percent is '%'. For example, 67% means 67/100.

Perfect Number is equal to the sum of its factors, except the largest (which is the number itself). So, 6 is a perfect number because 1 + 2 + 3 = 6. The next perfect number is 28, since 1 + 2 + 4 + 7 + 14 = 28.

Plus is the name given to the '+' sign.

Positive Numbers are greater than zero. Examples are 15, 7/10, 5.9, 31%.

Power is a number, also called the index, telling us how many of the same factor are multiplied together. For example, 7 x 7 x 7 x 7 can be shortened to 7^4 where 4 is the power. We read this as 7 to the power 4.

Prime Factors are factors which are also prime numbers. For example, the prime factors of 12 are 2 and 3.

Prime Number is popularly thought of as 'a number that nothing goes into' but more accurately is a number having exactly two factors, 1 and the number itself. The first four primes are 2, 3, 5 and 7. 1 is not a prime number because it has only one factor. Depending on the context, we can call these numbers 'primes', 'prime numbers' or 'prime factors'.

Product is the result of multiplying two or more numbers together. So the product of 2, 4 and 7 is 2 x 4 x 7 which is 56.

Proper Factors of a number are all its factors except 1 and the number itself. Only a composite number can have proper factors.

Quotient is the number of times a divisor can be subtracted from a dividend. What is then left is the remainder. For example, you can subtract 2 from 9 four times, so 4 is the quotient.

Glossary

Reciprocal of a number is 1 divided by that number. So the reciprocal of 4 is 1/4, and the reciprocal of 1/4 is 1 ÷ 1/4 which takes you back to 4. Any number multiplied by its reciprocal gives 1. The reciprocal of any positive whole number is a unit fraction. You obtain the reciprocal of a fraction by exchanging the numerator and denominator.

Rectangular Numbers are the composite numbers. They can be represented as a rectangular arrangement of dots where both the length and breadth are greater than 1.

Relatively Prime   Numbers having no common factor other than 1 are said to be relatively prime, sometimes called co-prime. For example, 9 and 14 are relatively prime.

Remainder   See divisor.

Semi-primes are numbers which have just two factors ignoring 1 and the number itself.

Sequence is a succession of numbers, or operations, in the order given. It is sometimes used interchangeably with series, but there is less sense of an underlying rule being applied.

Square number, or perfect square, is what you get when you multiply a number by itself. So 16 is a square number, because it is 4 x 4. This can be shortened to 4^2 which is read as four squared or four to the power two.

Square Root is a number which has been multiplied by itself. Thus 3 is the square root of 9 because 3 x 3 = 9. The square root sign is √, so √25 = 5 and √144 = 12.

Subtraction is the process of finding the difference between two numbers. It is indicated by the '–' sign, which is read as 'minus'. For example, 26 – 9 = 17. Addition is the opposite process.

Sum   This is the result of adding two or more numbers. So the sum, or total, of 28 and 19 is 47. The word is also used to describe other calculations, such as a division sum.

Total is the result of adding numbers together. Sum means the same thing.

Triangular Numbers are the numbers in the sequence 1, 3, 6, 10, 15 and so on, where each term can be arranged in triangular form.

The difference between successive triangular numbers is the sequence 2, 3, 4, 5, ...

Triple   Three of something.

Twin Primes are pairs of primes which differ by 2. For example 17 &19, 29 & 31.

Unit Fractions have 1 as their numerator.   For example, 1/8 and 1/5. Apart from 2/3, this is how the ancient Egyptians wrote their fractions. So they expressed 3/8 as 1/4 + 1/8.

Whole Numbers are the numbers 0, 1, 2, 3, 4 and so on.

Zero is represented by the symbol '0', also called nought or nothing. It is the first whole counting number.

If you have enjoyed this book there may be other Tarquin books which would interest you, including 'Cross Numbers' by John Parker and 'Number-Cell Challenges' by Wilson Ransome. Tarquin books are available from bookshops, toy shops and gift shops or in case of difficulty, directly by post from the publishers.
For an up-to-date catalogue please write to Tarquin Publications, Stradbroke, Diss, Norfolk, IP21 5JP, England. Alternatively, see us on the Internet at http://www.tarquin-books.demon.co.uk